It's Not *JUST* a Website...

The Small Business Owner's Blueprint for
Taking Your Business Online

Joanna! Great to see you again! Chris

Praise for *It's Not Just a Website*

"Simply outstanding! As a Business Coach I have worked with over 300 small businesses over the past 15 years. In all my experience one thing stands clear, not all websites are the same. And in today's economy a website is no longer a must-have. A website is an integral component of a successful business. Without it, you're losing business to the competition – every minute of every day.

It's Not Just a Website is a raw, honest look at the power (and traps) of growing your business online. It's packed with tips, strategies, and contains some of the best resources available. These tools are powerful and will make your business grow, it's that simple. This is hands down the best guide for small business owners I've ever seen."

- Jason Drees

Business Coach & Peak Performance Expert

JDR Coaching

• • •

"An incredibly honest, insightful and practical book that walks any business owner through WHY it's important to have an online presence then HOW to create, maintain and monetize that presence. Every small business owner must read this book, I personally will be reading and re-reading chapters 4, 5 & 6 as I work to drive more traffic to my website."

-Bibi Goldstein
Owner of Buying Time, LLC and Infusionsoft Expert

• • •

"I have managed online marketing campaigns for small businesses to huge Fortune 500 companies and having a great website is crucial in helping you monetize your website. Nearly all of the small business clients who I have worked with struggled and wasted THOUSANDS of dollars before they found me. *It's Not JUST a Website* will help you avoid the pitfalls that many small business owners make when taking their business online and it's a must-read for any first-time business owner.

Chris Martinez takes a very authentic no-nonsense approach. You can tell that he's the one with the knowledge as his personality and voice comes through his writing. Invest in yourself and read this book!"

- Wei Houng
Internet Marketing Expert, DigiWei.com

• • •

"I have been a friend of Chris Martinez for 28 years and have worked with him for the last 10 years on different marketing opportunities. Chris keeps it simple, to the point, and either embraces your ideas or tells you to move on. He uses the KISS approach and gets results. We worked on a plan to market potential sellers of businesses and real properties, after a short time he figured out a way to streamline the costs and I got results. This book is a clear model of Chris's approach to marketing and gives you REAL LIFE examples of what works and what does not when taking your small business online. I failed a lot and wasted more money, but in the end I came out on top because when Chris is involved failure is not an option. Email marketing is absolutely something that you can do yourself and it is something that you SHOULD do yourself and this is just one thing that Chris shows you how to do in this amazing book. One of my favorite lines that Chris writes is:

"Every single day that you wait to grow your business online, is a day that your competition is leaving you further and further in the dust, and one day they'll have moved so far ahead that you'll simply disappear".

I've been in business for over 30 years and this could not be more true. Read and learn this material immediately!! Do it now, don't wait."

- Brian Maginnis
Serial Entrepreneur and Founder of American Rentals

• • •

"As the Executive Director for SoCal BNI (Business Network International) in Orange County, Long Beach and the South Bay I have the opportunity to interact with over a 1000 small business owners on a regular basis. However, before I purchased the 3 BNI franchises I also ran my own insurance agency so I know first-hand just how difficult it can be to successfully get your small business online. Chris Martinez absolutely "nails it" when he breaks down the must-do's for any small business owner to be successful online. From creating your business website, to getting people to your site, to utilizing the latest online tools that help you generate real revenue, Chris breaks it down into simple steps that anyone can understand. This is THE MUST-READ BOOK for any small business that wants to get online today."

- Jenni Nering

Executive Director, SoCal BNI (Business Network International)

• • •

"Chris is passionate about safely escorting small business owners into the wilderness of the on-line world. I wish I had read a book like this before launching my first website because I could have saved myself thousands in trial and error!"

- Scott Martin

Founder of The Living Christmas Company and Winner of ABC's Shark Tank

• • •

"As the economy continues to move from brick and mortar to online, small business owners face an intimidating task of mastering all of the moving parts of the internet while still trying to get and keep clients and make payroll. In his easy-to-follow guide, Chris Martinez keeps it simple and direct in creating a basic blueprint ANY small business owner could pick up, read, and feel (perhaps for the first time) confident in either making a successful jump into the online economy, or upgrading their online presence."

- Timothy Ringgold

Co-Founder of Empower-U-Academy.com

• • •

"*It's Not Just a Website* is more than a book about website fundamentals. You really learn from a business owner's point of view how a website can dramatically grow your business. Regardless of the industry you are in, Chris Martinez presents an easy to read, no nonsense approach on how to establish your presence in the cluttered world of online marketing. This is a must read for any small business owner."

- Jeremy Mendoza

President, Annex Performance Solutions

• • •

"This book is like the small business owner's bible for getting online. I particularly enjoyed the chapter on Social Media and the impact that Social Media will have on our future. It is amazing at how sites like Facebook are changing our economy and I've personally implemented some of Chris's social media marketing into my new business. If you're a newbie to the online world and don't "speak geek" then this book is perfect for you. The language is simple to understand and you'll walk away with a clear understanding of how to take your business to the next level using the internet."

- Travis Winn

CEO, Trimr

• • •

"I became Mayor of Madera, CA because I've always felt a strong desire to serve and to better my community. Furthermore, as a business owner I'm a huge believer that entrepreneurship is a big opportunity for people to achieve their business and life goals. I've been blessed to have people like Chris Martinez to teach me about ways to grow my business online. If you want to take your bricks and mortar business online there are so many potential mistakes that you can make, but Chris's advice helps keep you on the right path. Read this book and I promise that you'll be able to successfully take your small business online. *It's Not JUST a Website* saves you time and money and you'll be learning from someone who is passionate about helping the "little guy" to come out on top!"

- Brett Frazier

Entrepreneur and Mayor of Madera, CA

It's Not ***JUST*** A Website

The Small Business Owner's Blueprint for Taking Your Business Online

By Chris Martinez

Foreword by Dr. Ivan Misner

IT'S NOT *JUST* A WEBSITE
THE SMALL BUSINESS OWNER'S BLUEPRINT FOR TAKING YOUR BUSINESS ONLINE

Contact the Author
www.WebsiteIn5Days.com

Cover Design and Book Layout created with love by
Eva Castorena

ISBN: 978-0-692-22133-4
Second Edition, 2014
Published in the USA

Dedicated To....

If you would have asked me 10 years ago if I would be in the website business I would have said "over my dead body". I never grew up as a "computer" person. I've been an athlete my entire life and I was misguided into thinking that computers were for geeks. Well, the times have a changed my friends. If computers are for geeks, then I am proud to say that I AM NOW A GEEK. (A very athletic geek, but a geek no less).

However, I'm not the kind of geek who can fix your computer like Jimmy Fallon's character "Nick Burn's" from *Saturday Night Live.* If your Windows XP crashed then you need to call the IT guy down the street. I'm an INTERNET guy, a website guy, and most importantly, I am the guy who can help small business owners to be successful online. I've honed my skills after years and years of floundering and wasting my own money and so you're learning from someone who has been in trenches.

Everything I know, I know because I have experienced it all on a first-hand basis and there have been so many people who have helped me along in my journey and to these people I would like to dedicate this book.

To John Marin, one of my first sales mentors who gave me my first Sales Management position while at Pitney Bowes when I was just 25-years-old, I want to thank you for giving me that opportunity. I know that you had to twist some upper management arms to get me that job and I learned more in that one year of management than all of my year's in college combined.

I also want to dedicate this book to the first Graphic Designer I ever hired, Christine Kenney. In 2007, I launched a print soccer magazine during the worst time in history to launch a print publication. Christine believed in my vision, worked her butt off for me, and went above and beyond the call of duty to design and layout all of our magazines. And she did it all for a fraction of what she should have been paid. She gave me my first introduction to the world of design and taught me that great designers are hard to come by.

Next, I can pretty much pinpoint my entire career in the online world to my good friend Amir Karkourti, Co-Owner of Surf Bros Teriyaki in San Diego, CA. Amir is not only one of the smartest, most inspirational entrepreneurs that I know, but he is literally the one who introduced me to WordPress and pushed me to figure it out on my own even though I had zero experience in web design. Without his encouragement, then none of this would be possible.

I want to also dedicate this book to Patricia Felan who was my manager when I took a job at ReachLocal where I would spend countless hours selling Pay-Per-Click advertising to small business owners. It is because of this job that I got to meet hundreds of business owners and came up with the concept for WebsiteIn5Days.com. Had

Patricia not given me this opportunity, I probably would never have made the connections that would inspire me to help start my business.

I want to dedicate this book to my Dad who never got to see me do any of these things. In December 2006 he was diagnosed with Pancreatic Cancer and he died a month later on January 10, 2007 at 12:10am, just three days before my 27th birthday. I think about him every single day and all I want in this world is to make him proud of me and to show him that I did something great with my life. Dad, you gave me more opportunities than anyone in this world and I just wish that you were here to share all of these new memories with me. You are the single greatest person that I have ever known and I hope that I can be half the person that you were.

No great business can succeed without a great team. To every single person who works for WebsiteIn5Days.com, I want to dedicate this book to you and your hard work. You are the backbone of our organization and you have helped hundreds of small business owners and helped us turn this dream into a reality. Thank you from the bottom of my heart.

To my dear friend Georgina Meza who lost her life in 2012 to cancer. You were taken much too soon and we miss you very much. And to her son Patrick and daughter Samantha, you have so many people who care about you. You two are an inspiration to me and seeing your strength propels me forward in my life and my business.

Lastly, I want to dedicate this book to every small business owner who has ever felt lost and confused by the internet. To the men and women who started their businesses with a dollar and a dream and are looking to take their business to that next level, this book is for you. This book is for the "little guy". The solopreneurs who eat, sleep, and breathe their business and just need a trusted advisor to help guide them to the promise land. This book is to help you maneuver through the minefield of BS that is out there. Everything in this book is to help you be successful and I hope that you can take my experiences and my knowledge to help you get to the next level in your life and your business.

Sincerely,

Chris Martinez
CEO/Co-Founder
WebsiteIn5Days.com

Foreword

I was honored when Chris Martinez asked me to write the foreword for this book, not only because he is a valued member of BNI®, the worldwide business networking organization I began back in 1985, but also because the subject matter of *It's Not Just a Website* holds particular resonance for me.

BNI started with a handful of members belonging to one tiny networking group in Southern California and it has now grown into an international organization comprised of over 100,000 members who pass billions of dollars in business to one another each year. Growing a business to these proportions has been no easy feat and I have faced more than my fair share of challenges. One of the biggest challenges I have faced throughout the years is centered on the very topic of this book—developing and maintaining an effective, ultimately profitable internet presence.

Internet technologies and digital media tools are in a constant state of change yet, as Chris so insightfully explains within these pages, the internet is not going away, it is not a fad, and if your customers cannot find you online, they will most definitely find one of your competitors. We live in a day and age that is immersed in ever-evolving technology and the fact is, small business owners across the globe in every industry imaginable are struggling with how to equip themselves to understand and embrace the internet revolution which is taking place. In order for businesses to survive and thrive in today's business climate, creating and maintaining an online presence is not an option – it is a necessity.

The good news is that you are holding in your very hands the ultimate resource for navigating the vast world of online media and internet tools. *It's Not Just a Website* is the essential guide to helping small business owners to get their business online and use the internet to develop visibility, credibility, and profitability in a powerfully effective way.

Chris Martinez is a straight-forward, no-nonsense author who uses his expansive knowledge, experience, and expertise in internet technology and online advertising to offer readers proven, results-driven advice accompanied by the simple steps necessary to create and maintain a highly effective web presence for their business. His immediately usable tips, tactics, and insights will enable your business to thrive online without wasting a lot of money on the unnecessary bells and whistles that many salespeople in the web development industry commonly try to persuade people to purchase.

In short, there is no better time than today to invest in the future of your business and to get online today. The world is changing faster every single day and, as this book states, "bringing your business online will not get easier over time." I urge you to immerse yourself in the invaluable wisdom and advice offered within these pages, to take notes, and to implement the practical strategies which abound from cover to cover. I have full confidence that if you do these things, you will successfully use the internet to take your business to new heights.

Ivan Misner, Ph.D., NY Times Bestselling Author and Founder of BNI®

Table of Contents

Introduction

Who I am

The world of business has changed more rapidly in the past 10 years than it has in the past 50. Despite all that has happened, we are still at the beginning of an industrial revolution that will change the landscape of business forever and there will be more millionaires (and billionaires) created in the coming years than ever imagined.

At one end of the spectrum we have the technological gurus and innovators who embrace technology and use it to build business empires. Then on the other end of the spectrum we have the small business owners who see the influx of technological change with fear, hesitation, and maybe even a little resentment. The funny thing is that there really is no difference between these two groups except for ATTITUDE.

Personally, I am fortunate enough to have gone from one end of the spectrum to the other, but not without taking my own "lumps". I grew up HATING computers. In fact, I've been known to throw a few keyboards across the room from time to time. I also started a print magazine in 2007, which was the worst time in history to start such a publication. I started this PRINT magazine despite some of my closest friends telling me that online was the way to go. What ensued was a fantastically enjoyed magazine that lost me a ton of money. Inevitably, had I not been so stubborn and had I taken my efforts online then I'm confident that my magazine idea would have succeeded.

The funny thing about failing in one endeavor is that you start to see the world a little differently and you see the opportunities that you did not see earlier. With my tail between my legs (and a giant hole in my bank account) I started to embrace this thing called "the internet". I taught myself web design by watching YouTube videos and I started to read everything I could find on the topic.

After building 15 or so websites on my own, my thirst for knowledge expanded and I started to learn about internet marketing and how I could get more people to the websites I've created. I went so far as to hire an internet marketing "coach" at $500 a month to teach me about how to drive traffic, get conversions, and make money.

Guess what happened? I failed a lot and wasted more money, but in the end I came out on top.

Why I'm writing this book

So you might be asking yourself, "what does this have to

do with me and how in the hell are you going to teach me about taking my small business online so I can make more money?!!""

Well, isn't that the million dollar question right there? How can I help YOU make money online?

Over the years I've had the pleasure of speaking with thousands of small business owners. Some were successful and some were not. Some were young and some were old. Some would listen to my advice and other would be stubborn.

The thing that they all had in common was that they all had a dream for where they wanted their business to go and they were all confused as hell as to how to incorporate this "internet thing" into their small business.

Through my business, WebsiteIn5Days.com, we set out to help the small business owner, the little guy, to maneuver this complicated technological landscape so that they could be successful. So far we've been doing a pretty darn good job at it as you can see if you check out some of the our client success stories on our website.

With that being said, the purpose of this book is to help another class of entrepreneurs and small business owners to be successful and achieve their dreams utilizing the power of the internet. Plain and simple.

Now this book is not for everyone. If you are too lazy and pig-headed to learn, then I cannot help you. If you think that you know it all, then I definitely cannot help you and I will happily refund your money right now for whatever you paid for this book. If you think that the internet is just a fad and that it will go away, then I suggest you find a beach to bury your head in the sand.

However if you see the opportunity for your small business to grow and you are willing to learn with an open mind then I highly encourage you to read every word of this book as your future success lies ahead.

Problems I see in the industry

I try to live an honest life and I've always been a very straight-talking guy so I'll be the first to tell you that there are a lot of companies in the online world who flat out just want to rip you off. They promise you some amazing website for $5000 that will solve all your business woes, bring you business, do the laundry, make the kid's lunch, and help you lose 25 lbs in two weeks. Pardon my French, but this is a load of bullshit!

If you've been in business for at least a year, I'm guessing that you've gotten at least 20 phone calls and countless emails from people claiming "they will get you on the first page of Google". More bullshit.

Like any new industry, there are those who can actually help you and there are the sharks who smell blood in the water and see you as the fat, elephant seal just bobbing in the ocean, waiting to be eaten alive. The only difference is that in the internet world the sharks look like life boats who are coming to your aid and you, for the most part, can't tell friend from foe.

It makes your job more difficult as a business owner, but this is just the world that we live in.

So by that logic, then how can you trust me?

You can't. But I'm not selling you anything. I do not want (or need) your money. My business was built on helping one client at a time and these clients continue to stay with us because we add value to their business. Lucky for me, many of them are so happy that they provide some pretty compelling testimonials about how we have helped them transform their businesses so I guess you don't need to trust me, but I would probably listen to my customers.

Anyway, there are tons of scammers out there who cannot help you, who over promise and under deliver, and who will rob you blind if you allow them to. However, do not think that you are off the hook and are not responsible if you ever get taken by these online gypsy bandits. In addition to teaching you how to make money online, I'll also teach you how to avoid the bad guys who just want to steal your money. It's up to you to use this knowledge to keep the bad guys at bay and to add more cash to your bank account.

The problems for most small business owners

I think that more so than the unscrupulous online marketing companies, the main reason why small businesses fail to make money online are the small business owners themselves.

Through my business I've interviewed thousands of small business owners and the thing that startles me most is how many of them are flat out LAZY and INCOMPETENT. It's amazing to me that some of these people even know how to tie their shoes.

Instead of embracing the new opportunities that are all around them, many of these nincompoops choose to put on their blinders and continue trotting down the same path that will ultimately lead them to the slaughter.

The world is changing every single day and if you do not adapt to the change then your fate will be extinction. The internet and all its opportunities are your friend and so I highly encourage you to have the bravery to forge a new path and grow your business in new ways.

Purpose of the book

After reading this book you will have an understanding of multiple ways to grow your small business online. You will probably also be offended at some point. I'm incredibly good at shaking you out of your comfort zone and pissing you off to the point that you take action. It always comes from a good place as my main objective is for you to be successful and to achieve your dreams as quickly as possible.

In 2007, I lost my dad to Pancreatic Cancer. In about a month I watched him wither away and eventually die in a hospital bed in the living room of our Palos Verdes, Ca condo. I saw the regret in his face about all the things that he didn't get to do and memories that he never got to create.

The one thing that you have in common with every other human being in the history of this planet is that you are going to die. Every single one of us has a clock that is expiring and unfortunately we never know when that will happen. It could be tomorrow, it could be 50 years from now, it could be in 20 minutes when you jump in your car to go down the street to get dinner. We flat out do not know.

I'm sure this is not revolutionary news to you, but the point that I want to emphasize is that if your life were to end today and you were no longer able to manage your business, what would happen? Would your business die along with you? Would it become a burden to your family and loved ones who would inevitably have to close everything down in your absence?

With the proper implementation of the principles in this book you can reach new heights of success faster than you ever imagined, but furthermore you can create a sustainable business that can live on long after you are gone if that is what you so wish to happen.

Grab a pen and paper and take notes as we are about to change your life.

CHAPTER ONE

If Your Customers Can't Find You Online, They Will Find Your Competitors

The world is online. The internet is not going away

I want you to stand up and raise your right hand and repeat after me

"I (state your name)
understand that the internet
is NOT going away
it is NOT a fad
and if my customers
cannot find ME online
that they will surely find
one of my competitors.
Furthermore
since my product or service
can help people
I have a MORAL OBLIGATION
to reach more customers
utilizing the internet"

I've forced roomfuls of people to take that oath and it's funny what happens when I do so. You can see the people in the room who are onboard and then on the flip side you can see the people out there who really do think that the internet is just a fad.

I recently saw a talk from a brilliant person and he said that the greatest inventions of the past 100 years were the INTERNET and the CELL PHONE.

So just go along with me and for the purpose of saving time, agree that the internet is not going away for a very, very long time.

There is a gap in small businesses being online

I have noticed that there is a distinct education gap among small business owners and it is inhibiting their ability to grow their businesses. This education gap has nothing to do with age, race, color, financial backing, but it has everything to do with ATTITUDE.

Your ATTITUDE towards being online is the single most important factor towards your success online. If you are like the "old" me and you are stubborn and refuse to embrace the internet then you will undoubtedly fail and potentially lose thousands of dollars.

However, if you accept that there are opportunities for your business to grow online then you have a much greater chance of success.

Let's just assume that you have been in business for 10 years or so and for whatever reason you haven't made the jump in taking your small business online. If you REALLY wanted to be online, which in this case simply means get a website, you could do so in literally 15 minutes. There are hundreds of companies out there that can help you set yourself up with a website and help you put that website online in less time than it takes to bake a cake.

In fact, we're inundated with advertisements from companies like Web.com, Wixx, GoDaddy, and even Vista Print who will help you create the ugliest website that you've ever seen for around $40 a month.

So since the opportunity to be online is all around us, then why is that that the latest study by Forbes Magazine shows that 52% of small businesses still don't have a website?!!!

52%!!!

I would make the argument that it is because 52% of small business owners need an attitude adjustment when it comes to the internet.

Like I had you stand up and swear an oath, "The internet

is not going away" and if your "customers cannot find you online, they will certainly find one of your competitors".

The internet is your friend and so change your attitude and realize that if you want to thrive in the new economy then you will need to embrace the fact that you will need to get your business online asa"f"p.

It's your opportunity to dominate your market if you act

A sales trainer once told me about how a potential client once told him "I've been doing (insert business here) for 25 years, why do I need to listen to you?" And he responded, "Sir all that tells me is that you could have been making the same mistakes for 25 years in a row."

I don't care if you've been in business for 20 years or 2 months, if you are smart and you market your business online you can 100% dominate your market.

The internet is the great equalizer. It gives those who are smart, brave, and persistent the opportunity to reach new customers in ways that has never been possible before.

Imagine that you have a street with 50,000 people walking down it and every single one of those people is looking to buy exactly what you sell. Would you not do everything possible to get on that street?!!

Even if there were just 50 people walking down the street looking for what I sell or even just 5 people I would do whatever I had to do to get on that street.

Well the internet is like that street and if you invest some time and money into building your online presence you can have an infinite stream of customers who can help you make all your dreams come true.

It will NOT get easier

If you think that you're going to wait for this whole internet thing to get easier for you, similar to how prices on Flat Screen TVs dropped significantly over time, I have some sobering news for you:

You're wrong and bringing your business online will NOT GET EASIER over time.

Every single day another business owner, or worse some giant corporate monster, is moving in on your turf. They are building a better website and marketing to your current and potential customers. Every single day that they are building their online presence and you are not is a day that you will never be able to get back and frankly at some point it will be impossible for you to catch them if they continue their marketing march.

It is almost like you are engaged in a battle for your survival on the internet. If you sit back and do nothing then the enemy's troops will continue to march towards your borders until it is too late and you are now at their mercy.

So consider my words a call to action! Get off your ass, arm yourself, and fight!

Now, I know that for most of you, the war analogies are a big turn-off. In fact, I'm not actually a zero-sum game kind of guy either, but this is reality especially if you are in a business that is fairly commoditized. If there is very little that differentiates your business from your competitor and price is really the only thing that your customers care about, then you are in extremely difficult waters. If your competitors are getting a strangle-hold on your online turf, then you need to act fast and you need to make some big marketing decisions to try and regain your position in the marketplace or you could find yourself in trouble very quickly.

What industries/people can stop reading since the internet won't help them

With all of my talk about the importance of you being

online, I will also tell you that there is a small segment of business owners who can pretty much stop reading right now. For these entrepreneurs, their industries are so difficult in the online world that I would not invest a lot of money in your online efforts outside of getting a simple website.

Insurance

Insurance is one of the industries that is unbelievably difficult to succeed via online marketing. You have a couple of major obstacles that stand in your way. First, if you are an agent under one of the major insurance carriers you undoubtedly have a very finicky compliance department that regulates everything from the mailers you send out to what you put on your website.

From a website standpoint, it is extremely difficult for you to get website copy approved that will actually increase business. Your compliance departments seem more concerned with not getting into trouble than with helping you get new customers and so even if you can drive a ton of traffic to your website you will not have the content that will get you business.

Second, insurance is one of the most expensive industries online and for the typical small agency you will need to invest thousands of dollars to get online, but the chances of making enough money to justify your investment might never happen.

Financial Planners/Advisors

Financial Planners are another group who find it difficult to succeed online. In terms of regulations and compliance, Financial Planners/Advisors are even MORE strictly monitored for what they can or cannot say and I'm not just talking about online. It's extremely hard for Financial Planners to get ANYTHING written in print or web.

Then on top of it, the small Financial Planner is also competing with the "Big Guys" who have unlimited budgets so it makes it very difficult for the individual planner to beat them. The Financial Planner might be able to get by with just their (bland) firm's website because the real marketing is going to be through networking, referrals, and hustling.

Firearms/Explosives Dealers

This section might seem like a joke, but Google (and I think Bing and Yahoo) have big restrictions when it comes to selling Firearms and/or Explosives. I doubt that this will ever change and most likely it will get even MORE difficult to conduct business over the internet for these folks. So basically, if you're in the business of making things go "BOOM!" then online isn't the place for you.

Fitness Products

Most people are surprised when they hear me say that fitness is one of the most competitive markets online, but it is incredibly competitive and it is unbelievable expensive between the months of November to March.

If you have a fitness product (DVDs are included in this category) you will 100% need a website to close sales. This is where most people will purchase your widget. BUT, you need an insanely large budget to drive traffic to the website OR you'll need a way to drive traffic to the website from other means. If you don't have deep pockets to invest in marketing then your chances for success are very slim.

CHAPTER TWO

What Is A Website Anyway?

Digital representation of your business

For the most part everyone has seen a website at one point in their life so I'm going to assume that I won't need to explain what the internet is and how a website is basically a group of files that people can access through this thing called the "internet".

What I want to talk about today is what a website is in relation to your small business. The best way I can define a website the way it relates to you is:

"A website is the digital representation of your business or brand."

A website is the way that your business interacts with customers and potential customers in the online space. Put another way, think of your website like your business's storefront on the internet.

If you think of your website as a physical storefront, would you take better care of it? Would you ensure that it is neat and tidy? Would you ensure that there is always a staff member there to assist your customers with any questions they might have? Would you make it SAFE for your customers to peruse your store? Would you update the look of your store to make it more appealing to the senses? Would you organize your products or services so that people can easily find what they need? And finally, would you engage in conversation with your customers to develop rapport with them and try to build a relationship so they become lifelong customers?

Hopefully you answered "Yes" to all of these questions regardless if you're an auto mechanic, an accountant, or a pet store. Most people would argue that all of these things are no-brainers when it comes to running a successful business in the physical world.

Well, I'm saying that your website should be no different and that you should treat it with the same, if not more importance and attention to detail, than your physical location.

Why? Because in the internet world, people are interacting with your business without you knowing about it! You don't know when they are on your site and you don't know what they are looking at so you don't have the opportunity to respond like you would if they were in front of your face.

That's why it's so important to make sure that your website is the best possible representation of your business.

Use your website as a Content Hub

Your website is also a content hub or in other words it is a place for you to demonstrate your expertise in your field. People these days want to work with an expert. Customers want to know that you really understand their needs and that you know all the ins and outs of what they want.

Your website is your opportunity to communicate your expertise to people. You can write articles, you can post testimonials of past clients, and the most powerful thing you can do is post VIDEO of you talking about how you can help people with their specific needs. Video is probably the #1 way to build a relationship with your clients and the best part is that these videos basically run on autopilot.

From a marketing perspective, search engines like Google love fresh, relevant content. Google rewards those who create fresh content on a consistent basis by giving you a higher ranking. So if you write articles or create videos that people like, then you have the potential to get more traffic to your website.

Your "channel" to broadcast to the world

I touched upon video in the last section which segues perfectly into treating your website like a "Channel" that you can use to broadcast to the world. Your website can be the most powerful media outlet that you have in your arsenal if you use it properly. With some hard work, your business can become the local authority and your website can become the media channel that people come to for the most up-to-date information.

If someone came to you and asked if you would like your own TV station for free, you would take it, right?

And you would take it because the TV station gives you the opportunity to reach thousands, if not millions of people, and you can leverage that media outlet to generate business for yourself.

That is exactly what your website can do for you. Treat it like a channel and it can bring you an unbelievable amount of business.

Get Personal and Tell Us YOUR Story

Building relationships is the foundation for a successful business in any industry. If you are unable to establish long-lasting business relationships then it is nearly impossible to run a long-term, sustainable business.

I'm sure you've heard the phrase "People buy from people they like and trust". Well, it is 100% true but I would like to say that "People buy ***over and over and over and over again*** from people they like and trust." That is really what the power of a strong relationship can do for you. It will create a life-long revenue stream for you.

A big thing that people overlook when trying to build relationships is telling their story. I see so many business owners who focus on the "What" in their business and they ignore the ***"WHY"***. I'm going to tell you that nobody is buying WHAT you do. They buy WHY you do it. I highly recommend that you go to YouTube and watch a TED talk by Simon Sinek. He gives statistical data that backs up this claim that people buy from you because of WHY you do it, not WHAT you do.

Your website is the perfect platform for you to talk about YOUR STORY and WHY you do what you do. Go deep and be vulnerable. Tell people the real reasons behind why you chose this crazy path of entrepreneurship. I promise that if you have the courage to be vulnerable then your ideal customers will be attracted to you like a moth to a flame.

For me, my "Why" it that I've always had a burning desire to create and to build something out of nothing. I also want to make my Dad proud and to show him that I made something of my life. And lastly, I want to build a successful business so that I can generate the income that allows me help to support the young man that I mentor and his sister who lost their mother to cancer in 2012.

All of these things make up my WHY and your customers will respond to that much better than you saying “We have the lowest prices in town”. AND they will stick with you for life.

CHAPTER THREE

The Basics of Building a Website

WordPress is King

This chapter is going to be more technical than most of the other chapters in this book so try to stay awake as I know that most of this stuff can get boring if you're not the technical type. I personally am not a very technical person so I won't use a lot of jargon and I'll do my best to keep it fun, but with that being said this is probably not the most exciting part of the book but it's one of the most necessary.

Now there are a lot of Do-It-Yourself website companies out there like Weebly, Wixx, GoDaddy, VistaPrint, etc., but if you're trying to build a serious online presence then there is no better Content Management System than WordPress.

Today, WordPress is the gold standard when it comes to building your website and some of the biggest companies in the world use WordPress.

What is WordPress?

I'm going to make this explanation super simplified. WordPress is basically a FREE platform that you can use to build your website on. WordPress was initially created as a platform for people to create a blog and over the past few years it has revolutionized the web design world and now it is the #1 platform on which to build websites.

One of the reasons it has become so popular, and I don't know if this is by luck or by design, is because it is incredibly search engine-friendly. Google LOVES WordPress and with Google's new love of fresh, relevant content WordPress has grown tremendously.

The other reason that WordPress has become the #1 platform to build a website on is because WordPress is FREE and what's known as "Open Source". Open Source means that anyone has access to it and can make changes, updates, revisions, and IMPROVEMENTS to WordPress as they see fit. You have an entire planet of incredibly talented designers and developers who are all using WordPress and are making it better. With that many people all working to make WordPress better it's no wonder that this platform has evolved so quickly.

The "1-Second Rule"

You may have heard that you have 3 seconds to capture someone's attention when they come to your website and if you can't get that person to recognize that you have the content they're looking for in that 3 seconds, then that person will leave and never come back.

I'm here to make this even more mind-blowing for you. Throw out the 3-second rule. We now live in a world of the "1-Second Rule". With all the distractions in our lives and the fact that we all want immediate gratification, I believe that you have just 1 second to get someone's attention once they get to your website.

That means that if someone goes to Google, types in "x", finds your website, and then clicks the link to your website, you only have ONE SECOND to show them that a) You have what they need, b) You are an expert, and c) You can help them RIGHT NOW.

You're probably wondering, "How in the world do I accomplish this in just one second?!" Well it's easier than you think and it starts with an easy navigation. You need to have a super, easy-to-read navigation that shows your services, your story, and gets people to click through on the website so they stay there. The navigation menu (this is what people often refer to as the "tabs" that link to the other pages on their website) should run at the top of your website and should run from left to right.

Your website needs to be an unbelievably sticky "net" to capture all your web traffic. If you've never done this before, then I highly recommend that you consult an expert web designer or marketer who understands this and can help you implement a design and a user experience that creates stickiness and helps you beat the 1-Second Rule.

Here's another tip to help you improve the stickiness of your website. Have pictures that capture attention. Great images can captivate your audience and entice them to stick around. Think about the last website that you went to and after you got there you decided to stick around. I'll give you a little hint: SEX SELLS. If you can somehow integrate images that have beautiful men and women then your stickiness will improve.

Another thing is that you don't have to reinvent the wheel when creating your website. As a small business owner you don't have to create some fancy, innovative website that the world has never seen before. Keep it simple. When we do websites for clients, we ask them to find a website that they like and that they would want to emulate. We never blatantly copy someone else's site, but if there are elements that are generating business for someone else and we can "swipe" that style, then that's exactly what we do. I encourage you to do the same.

Lastly, incorporate a video on your website. The statistics show that a website video will get viewed, on average, for 2 MINUTES!! 2 minutes versus 1 sec. That is unbelievably powerful and so if you want to overcome the 1-Second Rule then by simply putting a quality video on your homepage you will increase your stickiness.

Make Your Contact Info Easy to Find

Another huge factor in creating a great website that generates revenue is ensuring that your contact information is easy to find on every single page. The best place to put this is in the upper right hand corner or the website.

People read right to left, top to bottom, so by putting your contact info in the upper right hand corner you are placing that important information in a great location that will definitely get noticed.

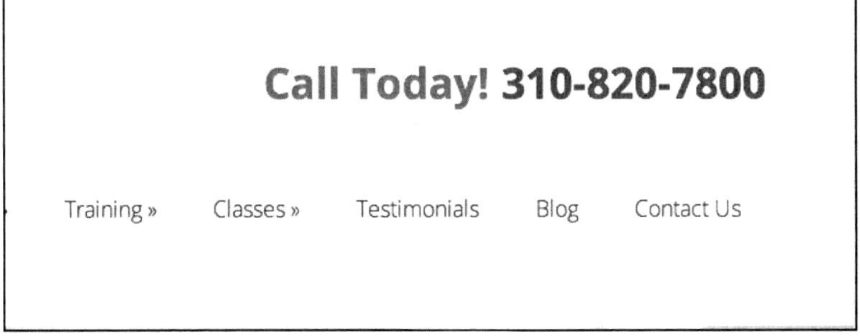

Call Today! 310-820-7800

Training » Classes » Testimonials Blog Contact Us

If you REALLY want to stand out, add a "Call to Action" next to the contact info. Have it say "Call Today" or "Free Consultation" or "Text Me for A Free Tour" or whatever.

People like to be told what to do when they come to your website so by simply putting a call to action in the upper right hand corner next to the contact information you can get a massive impact on your business.

It seems simple, but trust me...it works.

Put Up a Blog

We talked about using your website as a "channel" and enabling a blog on your WordPress website is the most important way for you to update content. From a marketing standpoint, the content on your "Pages" is not nearly as important as the content in your WordPress blog "Posts".

You should be updating your blog at least once a week, but if that's too much shoot for twice a month. Do it consistently and you'll be amazed at how quickly you can add more pages to your site and all of these pages give you another opportunity to be found online by potential customers.

The blog is also a way for you to position yourself as an expert in your industry. Here's a little hint: You don't even need to create your own content to be an expert. Look at Oprah. She doesn't really "create" content. She is like the world's most successful syndicator of other people's content. You too can become an expert in your field by syndicating awesome content that other people have created.

Just make sure that you "Source" and link to all the articles that you syndicate!

Use Video

I think this is probably the third time that I've talked about video and it's for good reason. As of 2014, if you have a video on YouTube, it is ***50 times*** more likely to be ranked on the 1st page of Google compared to an article. Furthermore, people spend on average, 20 minutes on YouTube when they go to YouTube *AND* YouTube is the #3 website on the planet.

If that doesn't get you excited about video then you need to smack your head against the wall. We are in the age of video and you need to jump on the bandwagon immediately. Once you post your video to YouTube, then you can easily post that video on your WordPress website. It is incredibly easy to do and your new video will keep people on your website longer and the longer they are on your site, the better chance you have to turn them into a customer. So put video on our website!!

Importance of Mobile

Did you know that there are more smartphones and tablets sold in the United States than laptop computers? Did you know that on average, 30% of all website traffic is coming from a mobile device and this number has been doubling every single year? And did you know that in some industries, the CURRENT mobile traffic to websites can be as high as 60%?!! Google even came out with their latest algorithm change, known as "Hummingbird", because of the increase in mobile traffic.

Mobile is an astronomically important aspect of your online presence that you need to be focused on. The first step is to ensure that your website has a "mobile-friendly" version which basically means that the screen size changes on a mobile device and tablet to make it easier for people to read. You can get a mobile plugin for your website from a company like Dudamobile (www.DudaMobile.com) or you can make sure that your website is utilizing what's called "Responsive Design".

(When we build our websites we make sure that all of the sites we build are utilizing "Responsive Design")

Mobile traffic will only continue to increase so make sure that your website has a mobile-friendly version or you will be losing out on potential business.

Google Analytics

Smart marketers are obsessed with analyzing data also known as "analytics". Lucky for you, Google has a free tool called "Google Analytics" that you can install on your website so that you can monitor all the data about the people that come to your website.

This is extremely important and I highly encourage you to install Google Analytics on your website as soon as you get your site built.

The Link for Google Analytics is www.google.com/analytics/ and again it is FREE. You'll just need a Gmail account (also free) to register.

With Google Analytics you can track things like:

- What location (cities, states, countries) that people are looking at your website from
- How many people are coming to your website and how many pages they are looking at
- What pages people are reading
- How long they are staying on your website
- How many people are looking at one page and leaving vs reading one page and then reading other pages
- How they are finding your website. Is it from google search? Another website that links to yours? An article you wrote? Etc.?

- How many people came to you from a mobile device.
- And much, much more!

Imagine this scenario. Let's say that you are a dog groomer in San Diego, CA and you have Google Analytics installed on your website. The data shows that 60% of your web traffic is coming from Yelp and then 20% of your web traffic is coming from a blog article you wrote about "Grooming Short-Hair Breed Dogs". Coincidentally, you've noticed that your sales have dramatically increased for people telling you they found you on Yelp and that you're noticing a huge increase in the number of Short-Hair breed dogs that you're grooming. Hmmmm.....very interesting!

So what would you do? How would you respond to this data and make an educated, business decision?

Here's what I would do. I would look at advertising on Yelp so that I could get my business in front of thousands *more* people. AND, I would also look at doing Google AdWords for "Short Hair breed" keywords (ex: "puggle grooming" "vizsla grooming" "pit bull grooming") and then I would also start doing a hell of a lot more blog articles on Short-Hair Breed dog grooming. Maybe even create a new VIDEO on grooming Short-Hair Breed dogs. Then I might also want to create articles on how to groom all kinds of different dog breeds. Long hair, curly hair, hypoallergenic, mixed breeds, etc., etc., etc.

See what you could do here?!!

Google Analytics can show you how to make great business decisions so therefore it can be one of the best tools in your marketing arsenal. So install Google Analytics on your website and get on your way to becoming a SMARTER and MORE INFORMED business owner!

CHAPTER FOUR

Got My Website...Now How Do I Get People To It?

The question I am always asked once the website is completed"

We've done TONS of websites for clients and 99.99999% of the time after we've built their site they ask "So how do I get people to my website?" It doesn't matter if when we started their website that they never intended to want to drive more traffic, once they see this awesome new website they decide that they want to drive as many customers to it as possible.

I'm going to give you some more tough love: Getting People to Your Website Is HARD!!

No matter what technique you choose to implement, it is going to be difficult and it's going to cost you time or money or both.

It costs money and/or time

Driving traffic to your website is all considered internet

marketing. There are lots of different ways to market your business and get traffic and I'm going to discuss each later on in this book.

What you need to know is that the world of internet marketing is filled with crooks who promise you everything, but really don't do crap. In my opinion, 90% of the internet marketers out there will rip you off.

How can this be you might ask? It's because the world of internet marketing is highly UNREGULATED. There is no entity that determines what "SEO" is and what is not. There is no organizing body that standardizes online marketing and so pretty much any company can say "We do Online Marketing" and since that term is not regulated it gives way to tons of scam artists.

How is the average person able to tell the difference between a scammer and the real deal? Unfortunately, YOU will never really be able to sniff out the real crooks 100% of the time. If you don't work in the digital world it is extremely difficult to find a reputable online marketer. However, later on in this book I'll show you some tactics to sniff out 80% of the bad guys.

I will tell you one thing: Online Marketing Will Cost You Time and Money or BOTH.

You can learn SEO on your own and you can rank your website, but that is going to cost you lots of time and probably a little bit of money. You can pay someone to market your website using SEO or run your PPC campaign and then obviously that is going to cost you money because you are paying them for their expertise.

Just like most things in life, you get what you pay for and the more competitive your industry, the more money you are going to invest to market your business online.

Think about how you get business today, incorporate that strategy

Often times when we talk to clients and they want to start "getting more people to their website" we shift the conversation about how they get business today. Most of our clients get their business from referral. So then I tell them then let's try to amplify the results of your referrals by putting more client testimonials on your website. Use what you're good at to continue to get base hits! Don't think about trying to hit homeruns! Do the little things great and you will win the game!

If you get a lot of business from The PennySaver or some other cheap ad, then look at driving traffic via Craigslist, or doing a mailer via Valpak.

Whatever way you choose to drive traffic, your new website can be the single best lead generator that you have.

Drive people to your website to claim a coupon, free report, or some kind of irresistible offer so that you can either 1) get them to call you to claim the offer or 2) get them to give you their name and email so you can follow up with them.

Even if you get them to the site, how do you capture them and nurture them?

In today's world, very few people will see an offer, call you, and then purchase. It just doesn't work like that anymore and Google actually did a study of this new phenomenon and they call it the Zero Moment of Truth (www.ZeroMomentofTruth.com).

People today will see your offer, check it out on your website, then look at your reviews, read about your bio on LinkedIn, look at your Facebook page, view your pictures on Instagram, check out your Yelp page, CHECK YOUR PRICES, and then after all that, they might pick up the phone and call you.

This is a fact and the data proves that this is what they do and they do it all in a matter of minutes because all of this information is readily available at their fingertips on the internet.

To be successful online you need to have a way to cultivate leads and nurture them. The easiest way to do this is to put a free or irresistible offer on your website that you give to people after they have given you their name, email, and sometimes phone number. This is often known as an "Opt-in Form" or a "Lead Capture form".

With your free or irresistible offer, you are having this conversation your web visitors:

"Hey Mr. Customer, this is Johnny Jones from ABC Company....I've got this incredibly valuable widget that I'm going to give you for FREE and this will help you make an educated decision or give you some value and in exchange, all you need to give me is your name and email. There's no obligation to buy anything from me and I just want to give this to you to show you that I'm a great guy and that I want to help you regardless of whether you give me your money or not."

Sign Up Today for a FREE 7-Day Trial Membership!

FIRST NAME:

LAST NAME:

PHONE:

EMAIL:

WHAT ARE YOU FITNESS INTERESTS?

- Group Fitness Classes
- Personal Training
- Youth Training
- Gym Membership

SPAM CHECK (Enter text below)

22868396

Type the text

Privacy & Terms

reCAPTCHA

SUBMIT

Once you get their name and email they are now on your list and you can start to cultivate that lead and build a relationship with that person over time. The more you do this, the more relationships you will build, and the more customers you will bring in!

CHAPTER FIVE

What Is This SEO Crap?!

What is SEO?

Most small business owners have heard of SEO (which stands for Search Engine Optimization), but they don't know hardly anything about how it works.

If you own a business today, you probably get at least 10 calls a month from random people promising to get you on the first page of Google. As I said in the last chapter, avoid these people like the plague. The majority of them are crooks and will rip you off.

My goal in this chapter is to just give you enough information to make you educated enough to sniff out the REALLY corrupt crooks. You will not be able to sniff out the crooks that are smart enough to disguise their crookery, but after reading this chapter you should be able to defend yourself against a modest crook.

So what is SEO? SEO is basically a number of techniques that you can implement on your website (known as On-Page SEO) and outside of your website (known as Off-Page SEO) to help you rank on Google for certain keywords.

A keyword is simply a phrase that someone types into a search engine, like Google, to get information. For example, if I want a haircut in a new city I'll go to Google and type in "Men's Haircut San Diego" and that phrase is my keyword.

There are thousands of keywords that your customers can type in to try and find you. If you implement SEO, you are trying to show up in the organic aka "White Part" of Google for specific keywords.

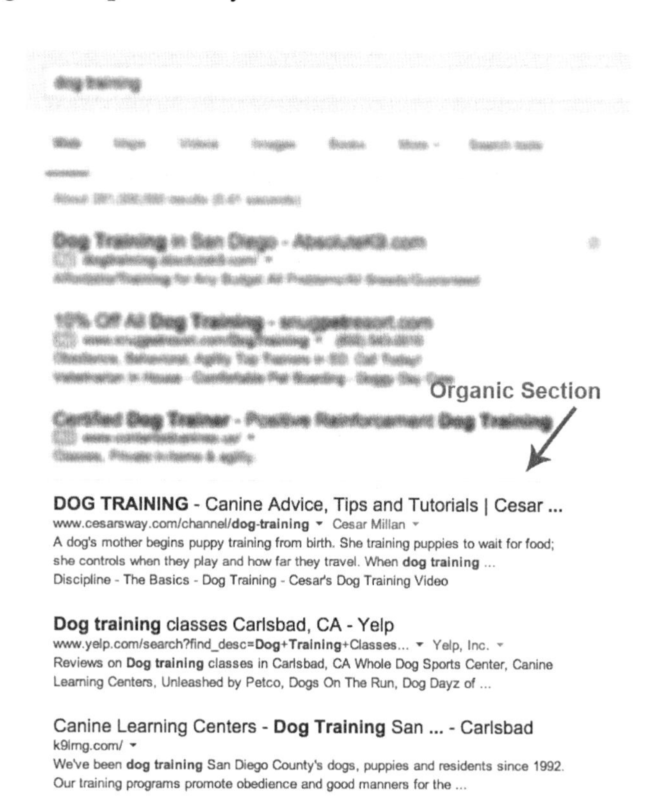

The best part about SEO is that nobody except for Google

knows how they really rank pages. It's a highly guarded, constantly changing algorithm that Google will never tell anyone about. However, there is a guide that you can download (it's called the "Google Search Engine Optimization Starter Guide") and it will give you an overview of all the things that you can do to optimize your website so that you can show up.

If you're the average business owner, this will take a lot of your time to learn. Years possibly. But it can be done. The legit people who do SEO spends years and years honing their craft and are constantly monitoring the trends to ensure that the sites they manage stay on top of the search engine pages.

However, you can definitely try to do SEO on your own if you have some time on your hands. I personally taught myself SEO when I was running a soccer blog and at one point we were generating about 20k hits per day to this site. So it can be done, but it's a lot of work.

How do I do it?

So let's say you don't have a lot of money to pay someone, but you have a lot of time to learn SEO yourself. I'm going to give you some shortcuts to help you with your own SEO. First off, you're going to want to build your website on WordPress and you're going to want to find a great SEO plugin to help you with your On-Page Optimization (I recommend **SEO Pressor** but **Yoast** is another great plugin that you can use, too). You're also going to want to identify some keywords that have low competition, but high traffic. Google has a free tool that you can use to identify good keywords. This tool is called the "**Google Keyword Planner Tool**". Talk about creative, right?

Once you finish your On-Page optimization you're going to want to implement your Off-Page Optimization strategy. You'll want to create "backlinks" to all your web pages and blog posts. Backlinks are basically links from other websites that point to your web pages and blog posts. It's like an authority thing. Think of each backlink like a "vote" that tells Google that you know what you're talking about and your content is relevant and valuable.

You'll also want to incorporate a Social Strategy that includes setting up your Google Plus page, getting followers, getting reviews, and then getting people to share your content on social media websites.

If you go online and subscribe to email lists from experts like Matt Cutts or Danny Sullivan then you can learn from them. You can also go to YouTube to get tutorial videos that will show you how to do all of this. Or of course you can find a company like mine, WebsiteIn5Days.com, and we can show you how to do all of this or do it all for you. However, don't expect immediate results with SEO. SEO is a long-term strategy and you will fail early on, but if you keep at it and keep getting better then you will eventually see results.

Who do I pay?

Again I repeat......Most Internet Marketers are Scam Artists!

Do NOT answer an email from a random company saying that they will get you on the first page of Google. I would recommend that you also not take a meeting with someone who calls you on the phone saying the same thing.

The best way to find someone to help you with your SEO is by referral. Find someone who used someone and make sure they got good results (Results = they are now ranking on the 1st page of Google for a few HIGH TRAFFIC keywords). Make sure that you check their references and that they are able to provide you examples of campaigns that they have run successfully.

By "High Traffic Keywords" I mean that they got your friend on the first page of Google for something like "Plumber San Francisco" or "Auto Repair Minneapolis". Anyone can get you ranked for an obscure keyword like "gluten free thin crust anchovy pizza in Peachtree City GA" because there is no competition. The good quality SEO companies will get you ranked for keywords that have a lot of competition.

If you have a good web designer then he or she probably has SEO experts that they refer to (In my company we have our own SEO Team). The trick is finding good quality people, who have reasonable prices, and that have the ability to take on more clients. Sometimes the best people don't have capacity for more clients so it helps if you have several options.

*If you have a proposal from an SEO company then you can always send it to us and we will evaluate it for free

Why does it take so long?

SEO is a long-term strategy so don't expect to get on the first page after one month. Don't even expect to get on the first page after 3 months. The majority of you will take between 6-12 months to show up on the first page of Google for one of your keywords.

Why?

There are a number of reasons, but I'll cover some of the most basic.

First, Google has these "spiders" that go out and crawl the web looking for relevant content. If you create a new website or a new blog post, it takes time for Google's "spiders" to find your page and then index the content. Indexing is basically taking the content that is on your website and then "filing" it so that the next time someone searches for something Google can go into its "filing cabinet" and pull up your relevant content. Considering that there are billions or trillions of websites out there, it takes a while for your site to get indexed.

Secondly, Google's #1 objective is to provide its searchers with the most relevant content possible. If you search for "dog training" and your search results pulls up and article about "cat training", then that search is not relevant to you. Google believes that since it's not what you were searching for that you will most likely never come back to Google so they basically lose you as a customer to some other search engine like Bing or Yahoo.

Since relevance is the most important thing to Google, then they are extremely particular as to what they rank on the first page. Therefore, your new post or website will have to gain "authority" before they put you on the first page. You will have to prove that your content is good, that people find your content relevant, that you have images that are eye-catching, and ultimately that you are serving people's needs.

The last reason that I'll give you as to why SEO takes so long is because every single market is extremely competitive and everyone is fighting for the same 10 spots on that first page. Let's take the plumbing industry for example. There are THOUSANDS of plumbers in Los Angeles and all of them would kill to be on the first page of Google when someone types in "Plumber Los Angeles". And so all of them are competing like mad to get one of those 10 spots, but no matter what there will only be 10 spots available so your chances of success are very, very small.

That is why a really good SEO company will help you rank for keywords that have HIGH TRAFFIC (i.e. lots of people searching for that keyword) and LOW COMPETITION. Your SEO expert should tell you how competitive your industry is and what keywords you should target. If they expect you to come up with the keywords or don't talk about the competitive nature of your industry then I would look for someone else to work with. This is usually a red flag that they are a scammer!

Why is it so expensive?

I'm going to give you a "BS Indicator" when speaking to anyone about SEO. Anyone who offers you SEO for less than $150 is full of crap. There might be a few people out there who are somehow able to do full-blown SEO for around that price, but for the most part anyone who is charging less than that is a scammer.

Typical SEO can range from $500 a month to $1000 a month to $10,000 a month depending on your industry and your location.

(You can go to VistaPrint or GoDaddy or some other template websites and they claim that they can do SEO for like $10-25 per month. This is one of the biggest lies in the internet marketing world and they are able to get away with it because SEO is not regulated in any way. Do NOT believe these big companies)

A Plastic Surgeon in Beverly Hills is going to spend WAY more than a Bookkeeper in Des Moines. The keyword "Plastic Surgeon Beverly Hills" is much more competitive than the keyword "Bookkeeper Des Moines" so therefore an SEO person is going to charge more because it requires more work to rank that keyword.

Depending on where you live, you can usually find someone reputable to do REAL SEO for between $500 to $900 per month. That includes a good On-Page and Off-Page strategy WITH a Social Media component. If they include content creation and articles you're probably looking more around $1k per month.

As with any form of marketing, this is an INVESTMENT in your business. If you decide to invest in SEO then know that sometimes your Return on Investment will take time. Make sure that you are getting monthly reports that show progress and make sure that your business is tracking phone calls and leads that are coming in from Google Searches. This is the only real way to track that your SEO is actually paying off.

CHAPTER SIX

Pay-Per-Click Might Be the Trick

What is PPC?

If you're looking for a form of internet marketing that can get you an immediate return on your investment then I highly recommend that you look at Pay-Per-Click advertising. Many people know this as Google AdWords.

Pay-Per-Click (PPC) is a way for you to advertise on Google or another search engine and your ads will show up for specific keywords that you choose, and then you pay for each click that comes to the website. If you signed up for PPC today, your ads could start showing up right away. It's the best way to show up on the first page of Google immediately.

Can I do it myself?

Absolutely you can do this yourself. You can actually call Google and they will set up your campaign for you and then you will be responsible for managing it.

I will warn you though. You can waste a TON of money (I have first-hand experience in this...before I learned about

PPC I tried to run my own campaign and spent about $3000 and got ZERO results) if you don't know what you're doing. Running a successful Google AdWords campaign is not easy and you will need to do a lot of research and you will need to constantly be tweaking the campaign in the beginning.

You will not get the most out of your campaign for a couple months. You should be able to make some money, but you won't really make the good money until you have a couple months of experience and have learned what keywords really generate leads and sales into your business.

There are two things you need to focus on when running your Google AdWords campaign.

First, you need to ensure that you have enough money to show your ad 24 hours a day. This is the biggest mistake I see people making. You pay every time someone clicks on your ad. If it costs $1.50 per click and your daily budget is $3 per day, then you're only getting 2 clicks per day. After you reach that $3 limit, your ad is done for the day and no other potential customers will see it! You are simply not giving yourself enough "swings at the plate". You need to have enough budget so that your ad can be seen throughout the entire day and so that you are getting exposure to all your potential clients.

The second thing you need to know is that your Return on Investment (ROI) is more important than your Click-Thru-

Rate (CTR). Clicks are awesome because they are driving people to your website, but in the end if those clicks aren't generating LEADS and SALES then it's all for nothing. You need to measure that your "clicks" are leading to sales. Google AdWords will show you what keywords are leading to clicks and then it is up to you to ask the client whether they clicked an ad or whether they found you organically. This is extremely important. You need to be able to measure ROI.

Many PPC companies and even Google will try to get you to invest more money in your PPC campaign because they got you tons of "clicks". Last time I checked, you can't deposit "clicks" into your bank account. It's your responsibility to make sure that those clicks are generating business. If they aren't, then change up your strategy.

Why does the cost per click vary?

PPC is like an auction and so basically the costs per click are based on what someone else is willing to pay for that click. Let's look at the top 3 spots AdWords spots on a website. These are at the top of the page and there are 3 spots.

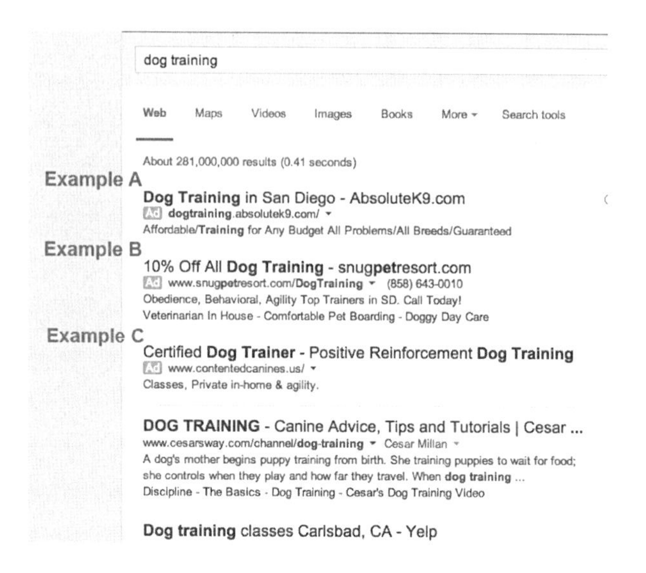

Let's just say that Company A is willing to pay $2.50 for that top spot. Then Company B has said they're willing to invest $2.40 for a spot so they get the 2nd Spot. And then Company C has said they will invest $2.30 for a spot and so they get the 3rd spot.

This is an incredibly simplified way to explain this, but basically you can see that you can buy your way into the party so to speak. The person who is willing to spend the most money can more or less have the top spot, hence the auction model.

Taking this into consideration, different markets and industries command different pricing because people are willing to spend more money. The Plastic Surgeon in Beverly Hills is willing to spend more money to advertise his boob jobs than the Pet Groomer in Des Moines is willing to spend for Chihuahua haircuts.

Can I pay someone to do this for me?

If I were a small business owner who doesn't know anything about Google AdWords, I would definitely pay someone to run this for me. In my own business, even though I know about Google AdWords I *STILL* pay someone else to run the campaigns for me. It is a way better use of your time to pay someone to run this than for you to learn it, fail at it, waste money, and then eventually figure it out and become proficient. Trust me. I've done this before on my own and when you look at time vs money, paying someone is a much smarter investment.

Most companies that run your Google AdWords campaign will charge anywhere between 10-25% of your budget and usually there is a setup fee that can range from $99-$499. That means that before you start your campaign you must pay them a setup fee and then once the campaign is running you pay them a percentage of your total monthly budget, which they call a "Management Fee".

For example, if your monthly budget to spend on AdWords is $1000 and the management fee is 25%, then you are going to pay the company who is doing the campaign $250 per month to manage the campaign for you. So $250 goes to the company, and $750 goes to your Google

AdWords advertising budget to spend on clicks.

That might seem like a lot but I promise you that the company you pay will be able to generate more results with that $750 than you would running your own campaign and with the full $1000 budget to spend fully on Google AdWords.

One thing to look out for when interviewing a PPC company (they also call themselves SEM companies) is that they provide you CALL TRACKING and MONTHLY REPORTS to show you how the campaign is performing. If they aren't showing you how many calls they generate and the keyword performance then walk away from that company and don't look back.

Will I need to do this forever?

CAN you run Google AdWords/PPC forever? Yes.

Should you run it forever? That's a question that you need to ask yourself as a business owner. Are you looking for new customers? Do you want to include this type of advertising as a way to generate new business? Is your existing campaign bringing you high-quality customers? If your answer is "yes" then I would run the campaigns for as long as they continue to generate revenue.
In almost any industry there will be changes, updates, or

new products that inevitably give you something new to sell. You can now use PPC as a way to show up for all these new advances in your field. So the short answer is this: Invest the money in PPC until it stops giving you a return on your investment.

Typically what we encourage people to do is to run SEO and PPC campaigns simultaneously. As your SEO is building, then your PPC can be used to generate immediate leads. Then once your SEO is up and performing, you can choose to scale back on your PPC or re-focus on keywords that are not yet ranking. However, if PPC is making you money then you can also make the argument to keep running it. Make that money!!

CHAPTER SEVEN

Social Media: The Future of the Internet

What is "Social Media"?

Everyone has heard about social media and the movie *The Social Network* about Mark Zuckerberg and the rise of Facebook is really would put the words "Social Media" into the mainstream consciousness.

Social media is basically a way for people to socialize online. You might be thinking to yourself "Who in the hell wants to socialize on the internet?!!....Go to a party geeks!" And if you're saying that then you have a good argument, however socializing on the internet is a little different than going to a party in the physical world.

Think of online socializing as simply a way to share ideas, thoughts, and entertainment. Hopefully you have a Facebook page so that this example makes sense. Think about how often you have gone to your Facebook page and check out the pictures that your friends have posted of their kids, or how you read an article that someone "Liked", or how you watched a YouTube video that someone else posted. That is social media. The sharing of information.

The same way that you would share "stuff" in a physical environment you can now do online and with people all over the world. As a business, this can be massively beneficial if you know how to engage your customers and potential customers in the right way.

Why should you care about Social Media?

If your ideal customers or business associates were at a party, would you not want an invitation to that party? Of course you would.

The main reason why you want to utilize social media is because all of your customers and potential customers are there. I promise that right now, as you read this book, you have customers and potential customers using social media. Not only that, they are sharing information about what movies they like, what sports they watch, what brands they support, where they ate lunch, and what businesses they recommend.

And it goes beyond Facebook. Social media is everywhere with many different social networks that you can be a part of. Obviously there is Twitter and Instagram, but Yelp! is also social media. Google Plus is also social media. YouTube could also be considered social media. LinkedIn is an incredibly powerful tool to connect with other business owners that often goes underutilized. Even your own website could be considered social media if people are sharing information on your site. Basically, the entire internet is now engaged in social media either directly or indirectly and so you should care about how social media impacts your customers and your business.

Can I make money using Social Media?

As I write this book, the monetization of social media (particularly Facebook) is EXPLODING. It's basically like the Wild West gold rush time for making money online and the ways to generate money by advertising on Facebook are changing and growing every day.

For example, let's say that you own a Yoga studio in Las Vegas. Your clients are typically moms who have school-aged children and these moms live within a 10-mile radius from your business. You can go on Facebook and run an ad campaign targeting that EXACT group!!

Let's say that you have a Martial Arts studio in Las Vegas and you want to run a campaign offering a special class for everyone who works at the Zappos Headquarters in town. You could run an advertising campaign to all the employees of Zappos who live in a 25 mile radius of Las Vegas!!

Let's say that you have a new offering that you want to give to existing clients. You could export your current customer list, match their phone numbers or email addresses to their Facebook profile, and then you could run ads to your existing customers, all on Facebook!!!

You can even upload your customer list into Facebook and then Facebook will create a "Lookalike" list of potential customers who look, like, and behave just like your existing customers. And you can then advertise to these people!

The targeting on Facebook is mind-blowing and all of the smart marketers are now using Facebook ads to win new clients. I highly recommend that you look at doing this to help you win new customers.

One thing that does NOT work and that I do NOT recommend is trying to sell people who have "Liked" your business page by blasting them with Facebook posts about buying stuff from you.

The posts are intended to engage your followers in a dialogue, not to try and sell them stuff. Think about it this way. Have you ever gone to a party and there is a guy there who is constantly trying to sell everyone something? You know. The cheesy guy who whips out his business card at any opportunity and tries to sell you his product or service when all you want to do is have a drink and some easy conversation. By the end of the night, the entire room is trying to avoid this guy like the plague.

Don't be that guy!!

Don't be the one on social media who is always trying to sell stuff at the party. Be the guy or girl who is the life of the party. The one with great stories. The one who knows all the cool stuff or hip trends. That is who you want to be when you post to your Facebook business page.

Can't my 22-year-old assistant do all my social media?

Just like any form of marketing, Social Media Marketing is a skill with a strategy behind it. Social Media Marketing is not just "posting stuff to Facebook". If you have a younger person in your office that happens to spend 22 hours a day on Facebook/Instagram/Twitter it **does not** make them qualified to run your business's social media page.

You need to have a solid strategy for social media and you need to integrate all your social media platforms so that you're conveying a consistent message that attracts the right customers to you.

There are usually social media courses in your area that you can sign up for or you can sign up one of your employees. Just remember, that coming up with a solid strategy and being consistent with your social media is key.

And most likely you will want to be the one calling those shots yourself. As with all major marketing decisions, make sure that you are the one setting the stage or you might not be happy with the results of your campaign!

Where is social media going?

I've always said that the online world is like "dog years". One year in the online world is like 7 years in the "real" world. This means that the things that are relevant today, most likely will be obsolete or will have evolved drastically by next year. Lots of people find this frustrating, but I think it keeps the world FUN.

We are presented with the opportunity to learn something new and exciting every single day. We get to learn new skills and impact people's lives in new and effective ways. And because of this we will never get bored!

With that being said, I think that social media will continue to blur the lines between the physical world and the digital world. We're already seeing this in some respects, but basically everything that takes place in the physical world will also have a simultaneous social component that takes place online and unites people from around the world.

As a business, there will be more ways for you to connect with your customers and build loyalty, but if you ignore the social media revolution then you run the risk of missing out on conversations about your business. You will need to have your ear to the ground so that you can listen and respond to all of your customers who engage with you online.

From an advertising standpoint, there will be more information collected about you, your likes and dislikes, relationship status, your stage in life, and countless other data points that advertisers will use to be able to predict with uncanny precision what you will want to do before you do it.

There was an article that came out recently that Facebook can now predict when a couple is about to change their Facebook status from "single" to "in a relationship" based on statistical data that they had gathered. This data included how many times the two people were in pictures together, how often they are "tagged" in posts together, and how often they write on each other Facebook profiles. Using the data from millions of Facebook profiles they are now able to pinpoint with pretty good accuracy when a couple will officially become a "couple".

As social media sites continue to collect data they will be able to predict more and more about your human behavior in the real world and so as business owners, we will now have opportunities to match our services with people at the exact moment that they want what we have to offer.
For example, a jewelry store will soon be able to serve

engagement ring ads to people who are highly-likely to become engaged this month because social media indicators will be able to predict that the guy will soon be getting on one knee. Social media data will show that the two people have been "In a Relationship" for 2 years, they recently took 3 vacations together, and in their age demographic they are very likely to get engaged. Based on the data, the statistical probability that this couple will get engaged is extremely high and so the jeweler will be able to target those two people with ads based on their likelihood of getting engaged.

A Mexican Restaurant will be able to serve ads to people who are highly-likely to want Mexican food this week because social media indicators show that this one person who lives within 5 miles of your Mexican restaurant likes to eat Mexican food once every 3 weeks and it's been 2 weeks since the last time he ate a burrito.

The list of scenarios like this is endless and as a business owner your opportunities to reach your customers at the perfect moment will soon be a reality.

Human beings are creatures of habit and since our lives have moved online there is an enormous amount of data that is now being collected that will allow advertisers to basically predict what we're going to want to do before we actually make that decision. It's going to be the most incredible social experiment in the history of humankind!!

I know that this is a lot to digest, especially if you are not already engaged in social media but I encourage you to learn it sooner rather than later. This social media landscape will only get more difficult and it will only get more challenging to make money in this space.

CHAPTER EIGHT

Banner Advertising: Digital Billboards

What is Banner Advertising?

Go to any major news and information website and you will see "banner ads". They are the little box-shaped or rectangle ads that are usually at the top and sides of any news website. I want you to think of these Online Banner Advertising just like billboard advertising.

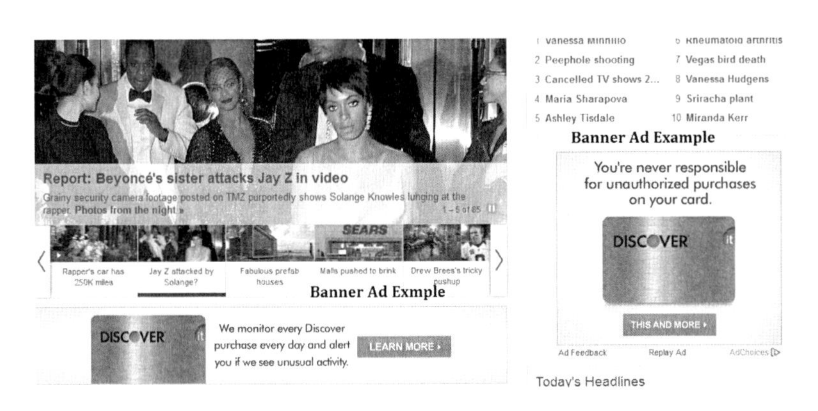

Banner Ad Exmple

Banner Ad Example

Banner ads are some of the oldest form of advertising on the internet and during the Internet boom of the late 1990s many companies made fortunes selling banner ads at what now are considered incredibly over-inflated prices.

Over the past decade, the effectiveness of banners ads has

diminished and so for the most part the only companies that use banner ads effectively are large brands with big branding budgets because nobody "clicks" on banner ads anymore.

There is even a phenomenon known as "Banner Blindness" where people now are so conditioned to seeing banner ads that we simply don't even notice them anymore.

I do not recommend buying banner advertising if you are a small business owner because it is highly unlikely that you will see any kind of return on your investment. Banner ads are also very hard to track in terms of how much traffic you are really generating from the ads themselves.

Some people will argue (these people being the people who sell banner ads) that a banner ad will inform someone about what you do, which will then prompt them to go to Google to research you, and then that person will eventually contact you. I personally think that those are a lot of hoops to get a client to your front door.

There are newer technologies like "Contextual Targeting" and "Behavioral Targeting" that can get your message in front of people who have already demonstrated an interest in what you sell or your business category, but I still don't think investing in this type of banner advertising is worth your money. Think about when is the last time you actually clicked a banner on a website? You probably can't, which proves my point exactly.

The only form of banner advertising that I do believe is beneficial for the small business owner is a thing called "Banner Retargeting".

Banner Retargeting is basically a way for you to keep "Top of Mind Awareness" with someone after they have come to your website and gone.

We've all experienced this before. I went to stay at the MGM Hotel in Las Vegas once and so I went to their website a couple weeks before the trip and then for the next two months I kept seeing ads on all these other websites for the MGM Hotel in Las Vegas. The reason this happened is that when I visited the MGM website, they placed a "cookie" on my browser, and then for the next 2 months whenever I visited certain websites I was shown the MGM banner ad.

Banner Retargeting can be very effective in keeping your brand in front of people and is a good investment for people who have products or services that have a long sales cycle. If you typically have to work with someone for 2-3 months before they become a customer then Banner Retargeting could be a great tool for you.

We've been using AdRoll.com for our banner retargeting and it has been very easy to set up and also very affordable. If you're interested in banner retargeting I would recommend that you give them a try. They also have a 2-

week free trial so there's no risk.

Where do I buy the ad space? What's a CPM?

You can buy banner advertising typically from any website that is out there, but the bigger the website, the more expensive it is to buy space, and the higher the minimum advertising spend is going to be.

If you live in a major metropolitan area, a great place to go to buy banner advertising on the bigger websites is through your local newspaper. Most of the local newspapers that are still in business have relationships with major banner ad networks (they call these "Display Networks") and so you can leverage their relationships to get featured on major sites like CNN, Yahoo, MSN, etc., for pretty reasonable rates.

(One thing that you should know when going through these Display Networks is that you won't get what is known as "Guaranteed Placement" meaning that your ads will go into a rotation and so they will be shown pretty much at random. That is why you are able to buy the ad space at a discount. Big companies who have huge budgets will buy "Guaranteed" space and pay a premium to be there. As the little guy who doesn't have $2k per day to invest in that kind of advertising, you will get the random ad placements, but you still get to be on those larger sites and could attract those customers.)

There are 3 standard advertising sizes when buying Banner Ads. You have the Rectangle, also known as the Box Ad or the 300x250. You have the Skyscraper which is a tall, narrow rectangle ad that runs down the side of a website and is also known as a 160x600. Then you have the Leaderboard which is a skinny rectangle that runs horizontally across a website at the very top, the middle, and the bottom and is also known as a 728x90.

Unlike Pay-Per-Click advertising where you pay for every "click" to an ad, Banner Advertising is sold by Impressions or basically how many times the banner is displayed on a web page. You buy banner ads on a Cost Per Thousand (CPM) basis, which means that for every 1000 times a banner is displayed you will generally pay $X. Cost per thousand can run anywhere from $1 CPM all the way up to $30 CPM depending on the website and the kind of targeting that you are doing.

Typically, the Banner Retargeting that we talked about in the last section will cost you around $10-$15 CPM. However AdRoll.com is offering it around $1-2 CPM which is a really good deal.

Banner advertising can get very expensive if you really want to get good saturation with your market. Typically, you want to have a minimum of 20k Impressions per day to really have an impact on your potential customers. If your Cost per Thousand (CPM) is $10 that equates to $200 in ad spend per day. From those 20k Impressions, you will likely generate 10 Clicks to your website. $200 spent for just 10 clicks to the website....That is a very high price to pay to get someone to your website!!

So basically stay away from Banner Advertising unless you have a very large budget and you are doing a branding campaign.

Where to get the banners created for cheap

If you do decide to do banner advertising (hopefully you're only doing Banner Retargeting) then the company you work with might offer to create your banners for free or if you're lucky enough to have in in-house graphic artist then you can have him or her do it for you.

If you're not lucky enough to have these options at your disposal then your other option is to go to Fiverr.com and get your banner done for just $5. You can find a ton of graphic designers who will design pretty much anything for you for only five bucks...hence the name Fiverr.com.

I've used this site a bunch for a multitude of things, but the banners I've gotten from them are decent. I would not say that they are fantastic, but then again you are only paying $5.

If you really want to get sneaky with your banner art and the website you're advertising on will allow you to do this, you can create your banner to look like an ARTICLE on that website.

Basically take a screenshot of the website you're going to be advertising on, then tell your artist to make your ad look like an article on that site. You'll have to create several headlines and story summaries that all relate to what you do, but I've used this technique in the past to generate a lot of traffic to my website. The only problem is that most major websites won't let you do this kind of thing. But it's worth a shot if you decide to do banner advertising.

CHAPTER NINE

Email Marketing: Your Biggest Money-Making Tool

What is email marketing?

Of all the strategies to drive traffic and get SALES, email marketing is by far my favorite and most successful. On average, for every $1 you invest in email marketing you will generate $43 in sales. Usually when I do my campaigns, I see a much higher return on my investment than that, though. Just last month I sent out an email campaign that grossed nearly 1000% ROI!!! Now THAT is some serious advertising returns.

Email marketing is basically what it sounds like. It is sending emails to people who have shown interest in what you do or who are potential customers. Some people call this SPAM, but if they do business with you or have Opted-In to a mailing list that you promoted then you have the right to email them.

Obviously you should respect people's wishes and if they don't want to be contacted by you then I would take them off my list, but that choice is yours.

Do I do this myself?

Email marketing is absolutely something that you can do yourself and it is something that you SHOULD do yourself. If you have your own list, which for most of you will be your existing customer database, you can upload that list to a service like Constant Contact, iContact, Vertical Response, or Get Response and then you can start email marketing to your list.

Each of these services have fancy templates that you can use to mail to your people, but I have found that I get the best response when I send out the most plain, boring-looking emails. Literally, just have text. No pictures, no fancy background. Just text. The other thing you want to do is make it so that the paragraphs are narrow.

Also, and this is VERY important, only write your sentences approximately 11 words across. Don't have long sentences that take up the entire screen. The statistics show that those long sentences confuse the reader and therefore your emails don't get read.

And don't have more than 5 lines in each paragraph.

I don't know WHY this is the case, but time and time again the results have proven that emails that are formatted like this generated the most leads and sales. So why reinvent the wheel? Go with what works!

What if I don't have a list?

So let's say that you are a new business or that you are an existing business and you have somehow managed not to collect email addresses from your customers and potential customers.

(If you are the second scenario, SHAME ON YOU! You basically hate money and deserve to be smacked! Your customer database is the most important asset in your business so make sure you have information, including email addresses, for all your current and past customers.)

There is a way to send out mass emails even if you don't have a list. You can go to companies that have giant email databases and you can rent their list. Basically they will send out an email to people on your behalf. You personally won't get access to the list itself, but if people contact you for more information then you obviously get to keep their information and can build your list that way.

I've seen these services from anywhere between $800 to $1500 for emailing to a 40k person list and the targeting is really, really good. Typically you can get a few hundred to a few thousand people to your website depending on how great your email offer is.

One thing you do NOT want to do is BUY A LIST. DO NOT BUY A LIST! This is against the law and you will have your email marketing account deactivated when you start getting a ton of spam complaints. So just don't do it!

Warning About Email Marketing

I wanted to issue you a warning about my favorite marketing tactic. The delivery of email marketing has recently been negatively impacted due to a new change by Gmail. Gmail implemented a new tab in their inbox that they call "Promotions" and it is there to simply syphon out all email marketing. Because of this, many email marketers are seeing their ROI impacted in a negative way.

As much as I love Email Marketing, this could be a signal of things to come. Much like Fax Blasts have disappeared since the Late 1990s, email marketing could also go the way of the dodo bird.

With this in mind, I encourage you to use email marketing today, but also start to incorporate other ways to market your business in case email marketing was to disappear.

CHAPTER 10

How to Monetize Your Website and Online Presence

So how do I make money with all this "stuff"?

I've thrown a TON of information at you in these past few chapters and I know that your head is probably spinning after all the acronyms and crazy "Internet Geek Speak" that I've given you, but I want you to pay close attention to this chapter because this is where I teach you how to leverage all this knowledge and MAKE MONEY.

By now, I've given you multiple ways to drive people to your website. I've also given you the keys on how to design your website to get leads and inquiries.

The key to monetizing your website now is to go out and IMPLEMENT. Make it happen! Try everything and see what sticks!

I'm going to give you the cold hard truth. Being successful on the internet is not for the faint of heart. You need to be brave and you need to be prepared to get your butt kicked and to lose some money. But think of it as an INVESTMENT in your future.

Just like in baseball, you are going to swing the bat 7 times out of 10 and you will strike out or not get on base. But if you can consistently hit the ball and get on base just THREE TIMES, then you will find yourself in the Hall of Fame.

Be prepared to FAIL

Zig Ziglar has a saying that "Anything worth doing, is worth doing poorly" meaning that if there is a big reward for some massive undertaking, then it is worth it to suck at it before you get good. Being successful online is no different.

You are going to get your butt kicked pretty much every day for the first year, especially if you are trying to do everything yourself. You're going to lose money, you're going to want to say that "This internet crap doesn't work", and you're going to want to throw in the towel and go back to your old ways. But I promise you that if you stick with it, you will find a recipe for success and you can make more money than you ever dreamed of.

The best part is that when you start TODAY you automatically get one step closer to dominating online AND your competitors who refuse to adapt to the digital world are now getting smaller and smaller in your rearview mirror as you move along down your

technological road to riches.
So don't be discouraged and keep on moving forward and by all means, if you can find a partner to help you make sense of the internet and help you avoid some of the pitfalls then you should leverage their knowledge to help you shorten that learning curve.

It will take money and time to be successful

Taking your small business from the offline world and building your digital dominance is not something that happens overnight. It will take time and money just like anything else.

You will invest and invest and invest and not see a return, but eventually you will start to see your money come back to you in exponential amounts.

That is not to say that you should just continue to plow along pouring money into things that aren't working. On the contrary, if you are not getting the results that you're looking for then you must change your approach. My recommendation is to always test the waters before you start any marketing initiative. Run a small test on a small sample size to see if you get the results that you want and test out multiple messages. You never know which message and which marketing strategy will end up being the winner for you.

There are a lot of people looking to rip you off

I've mentioned this several times, but the internet world is filled with people who are looking to make a buck off of your naiveté. It's sad and disgusting, but it is a fact and unfortunately, I would say that overall there are probably more BAD Web Designers, BAD Internet Marketers, and BAD Online Gurus than there are good ones.

It is extremely tough for someone like yourself to be able to decipher the good guys from the bad guys so it helps if you have someone in the technology arena that you can turn to for answers. In my company, we do this all the time. We speak with people who we know will never be our clients because our goal is to help them.

We believe in the philosophy of "Givers Gain" meaning that the more you give to the world, the more you will receive in return.

So therefore my offer still stands. If you have any Online Marketing proposal or initiative that you want me to look at, feel free to call me and we will go through it together. Free of charge. No obligation and no questions asked.

Expand your business into new areas with the Entrepreneur's mindset

If you really put forth the effort and embrace the internet, it will open up new revenue streams for your business that you never knew existed. The internet will bring more people into your life and into your business and those people will share with you what they need and just can't seem to find. This will be your opportunity to create new products and services that can help these people.

You might say that you never intended to be in this business or that business, but if you are a true entrepreneur then you will seize these opportunities that come your way. Nearly every single successful company in history has experienced a time where they had to "pivot" and change their direction. They initially started out going in one direction with their business and providing one service, but then the market responded and told them "Hey, you should go this way" and these companies adapted and became wildly successful.

The internet can help you uncover these new pivot points in your business and I don't care if you're 19 or 90, you have the ability to reinvent yourself and your business when this happens! So I challenge you to have the courage to push forward into the unknown and welcome the digital changes that could just be the best things that ever happened to you.

Never stop learning

If there is one joy that the internet world has brought me it is that I am constantly being challenged and am learning new skills every single day. There is literally not a single day that goes by where I am not being taught a new skill from something I found online.

I think that this is one of the keys to a happy life. Developing a love of learning is something that will keep you excited to get up every morning and will empower you to rise above those who are too lazy to continue this education.

The things that I've taught you in this book are just a fraction of what is out there, and frankly in the next 18 months most of what I teach will be irrelevant. Some people might find that annoying, but I think it's the best part about growing your business online. There is always something new to learn and always some new hill to climb. That's what makes this business so fun!

With that being said, I hope that you will embrace the opportunity to further your education beyond this book and that you will leverage that new-found knowledge to grow your business to new heights so you can achieve all the things that you want in this world.

Conclusion

I consider myself to be a great "student of successful people". I follow them, read about them, watch TV shows and interviews on them, and over time I have identified patterns that link them together, despite their different industries and backgrounds.

So why is it that some people reach the highest heights of success and most others are content with being somewhere in the middle?

I think that it has a lot to do with COURAGE. You might have heard that in emergency situations like a natural disaster, fire, or terrorist threat that 90% of people simply "freeze". They are unable to move, unable to think or react, and they are essentially waiting for someone to save them. Then you have the 10% of people who rush into burning buildings and do unthinkable things to try and help people they have never met.

I believe that the 10% of those who have the courage to ACT in business are the ones who are most-likely to succeed in this world and I definitely see it in the digitization of small business.

The winners in this game will ultimately be the brave ones who ACT and don't sit on the sidelines waiting for someone else to make the first move. If your mindset has always been "Well, I'll see how it works out for that guy and then I'll give it a try", then I'm sorry to say that you're only making things harder on yourself.

Every single day that you wait to grow your business online, is a day that your competition is leaving you further and further in the dust, and one day they'll have moved so far ahead that you'll simply disappear.

This book is a call to arms for small business owners. I've seen so many people waste opportunities to grow their business because they are fearful of change. Regardless of whether your business is thriving or if you are days away from closing your doors, the key to your success is ACTION.

You were born with the capacity to achieve anything and everything that you want in this world. Your success is limited only by your imagination and if that little voice in your head has been telling you that "You're not ready" or "We'll do that next year" then it's time for you to man or woman-up and FIGHT. The universe will reward your action.

I'm so incredibly proud that you gave me the opportunity to share my knowledge with you in this book. There is nothing that makes me happier than helping someone else. It's more rewarding than money or vacations or anything like that. I've given you tools that can help you take your business to new heights and hopefully you will see that there are even more things to learn that will help you achieve your goals. So now there is only one thing left to say:

The time is now. You have no excuses. Put your head down and get to work.

Glossary

Banner Retargeting- A way to place a "cookie" on someone's browser once they come to your website and then once that person leaves your website they will start to see banner ads for your business on websites all across the internet. This is the best form of Banner Advertising for small business owners and helps you create Top-of-Mind-Awareness with potential customers.

Behavioral Targeting- This is a form of banner advertising in which a person's search behaviors are tracked and then they are served banner ads that are related to their searching. An example of this is that a woman is searching online for information about planning a wedding. Soon after she starts to see ads for wedding venues, wedding flowers, and all things related to weddings. The Display Networks were able to track her behavior and then target her relevant advertisements.

Contextual Targeting- This refers to a form of banner advertising where your banners are displayed on websites that are similar in their content to your business. For example, if you are a roofer you can use Contextual Targeting to display banner ads on websites that are related to home improvement.

Conversion- In the online world a "conversion" is recognized as either a lead or a sale generated on your website.

Conversion-Based Optimization- This is a term used when driving traffic to your website and allocating more budget to the keywords that generate more Leads or Sales on your website. For example, if you are running a PPC campaign and the keyword "Dog Trainer" is generating 5 Leads or Sales every month, whereas the keyword "Dog Training" is not generating anything, you would use Conversion-Based Optimization to allocate more budget to the keyword "Dog Trainer" since that is generating more results for you.

Cost Per Lead- this is how much money you invest to get a lead into your business. If you spend $100 for 5 leads, then your Cost Per Lead is $20 per lead. (I.e. Total Spend / Total # of Leads = Cost Per Lead)

CPC (Cost Per Click) - This is how much you spend for each click to your website. You'll always want to look at your CPC as it will factor into your cost for acquiring a new customer.

CPM (Cost Per Thousand) - The cost metric for buying Banner Advertising. It is the cost for every 1000 views on a given website. There is no guarantee that each web visitor will see the and the impressions could be from the same person. The formula for calculating CPM: Cost / (Total Impressions / 1000)

CRM (Customer Relationship Management) - An online

software tool that you can use to manage customers. You can use this tool to track leads, proposal, existing accounts, and the lead capturing tools within the CRM can be integrated with your website.

CPA (Cost per Acquisition) - Another way that you can pay for advertising. In this model, you only pay for Leads or Conversions, and is an optimal way to buy advertising (if you can get it).

Display Network- A Display Network refers to a massive group of websites that an online advertising agency can use to place your banner advertisements. There are many Display Networks out there and many of them place ads on similar websites so there is a lot of overlap. Although Google has the largest Search network, the largest Display Network in the world is Yahoo!.

Google Analytics- This is the free analytics software that you can install on your website to help you track all of the traffic to your website. This is a very powerful tool that you can use to make educated decisions about the updates, changes, and traffic strategies that you can implement on your website.

Landing Page- A one-page website that is used in conjunction with a targeted advertising campaign. The sole purpose of this webpage is to collect leads via an Opt-In form, emails, or phone calls. This website usually incorporates promoting a single product or service and usually includes an irresistible free offer in exchange for the viewer's name, email, and sometimes phone number.

Lead Conversion Form- This is a sign-up form that you can put on your website to collect your web traffic's contact information. You can set up the form to collect a wide range of data, however almost all Lead Conversion Forms collect the person's email address.

Navigation Menu- Also called the "Nav Menu", this is the list of pages that are listed on the homepage of your website. Often times these are referred to as the "tabs" across the homepage.

Opt-In Form- The same as a "Lead Conversion Form", this is a form that you can put on your website with the goal of collecting contact information from your web viewers and turning them into leads.

Organic Traffic- This refers to people who come to your website from the Search Engine Results Pages on a search engine, not the paid listings. This is sometimes referred to as traffic derived from SEO.

PPC (Pay-Per-Click) - Is commonly associated with a form of advertising on search engines like Google (On Google PPC is often referred to as "Google AdWords"). PPC is the method that search engines use to make the majority of their advertising revenue.

ROI (Return on Investment) - The amount of Revenue or Profit generated from a marketing initiative. Ultimately profit is the most important factor when looking at ROI, but advertising reps will usually only focus on the Revenue generated from an advertising campaign. You'll want to look at both when looking at your true ROI.

SEM (Search Engine Marketing) - A way of promoting your website and getting people to your website by purchasing traffic. SEM can refer to Pay Per Click advertising, purchasing Banner ads, and other paid traffic channels. One of the biggest advantages of SEM is that you are able to drive traffic to the website right away.

SEO (Search Engine Optimization) - The process of affecting the visibility of a website or a web page in a search engine's "natural" or un-paid ("organic") search results. In general, the higher ranked on the search results page you are, and more frequently a site appears in the search results list, the more visitors you will get from the search engine's users. Seeing results from your SEO so that you "show up on the first page of Google" can sometime times several months or even YEARS depending on the competitiveness of your market.

Recommended Reading and Resources

Glazer-Kennedy Insider's Circle- I am a big fan of Dan Kennedy, the master of Direct-Response marketing. I have been a member of his group "Glazer-Kennedy Insider's Circle" for several years (www.DanKennedy.com) and I highly recommend that you join if you want to learn how to market your small business. Some of my favorite books that Dan has written are *No B.S Marketing to the Affluent* and *No B.S. Sales Success in the New Economy.* Dan's books all have the "No B.S." title and I recommend them all, but those are some of my favorites. I also recommend his business partner's book *Outrageous Marketing that is Outrageously Successful* as this is a fantastic book that teaches you some "unconventional" methods to attract customers.

The Search: How Google and Its Rivals Rewrote the Rules of Business and Transformed Our Culture- Written by John Battelle, this book was given to me when I started working for ReachLocal Inc. and it is possibly the single greatest book to understanding how Search Engines, like Google, make money and the history behind how Google rose to power and changed the world.

The E- Myth- A friend of mine gave me this book and it is the single greatest business book for first-time business owners. The author, Michael E. Gerber, shows exactly what it takes to run a successful small business as well as the "mind shift" that one must undergo when going from employee to owner. I highly recommend that every small business owner read this book, but especially the first-time business owner. Simply by reading this book you will save time and money.

David and Goliath- Malcolm Gladwell is best-known for his books *The Tipping Point* and *Blink*, and this newest title is especially relevant for small business owners. *David and Goliath* pinpoint strategies that "the little guy" can use to beat out stronger, better equipped competitors. It's a pretty fast read and you can skip some chapters, but in the end you'll have a great idea as to how you can beat out the "Goliaths" in your industry.

AdRoll.com- This is the website that we're currently using for our banner retargeting. They have a great support team to help you get set up and their prices are really cheap at around $1-$2 cpm. They don't help you with the artwork for your ads, but just use Fiverr.com if you don't have an in-house graphic designer.

Fiverr.com- This is an awesome website that you can use to get pretty much anything done for $5! I recommend using Fiverr.com for getting banner ads created, but you can search through the services and see what else you might be able to get created for just 5 bucks.

LeadsPages.net- Need a landing page quickly and for cheap? Go to this website to search through dozens of templates that you can select for your landing pages. We've used them in our business and I highly recommend them if you need a quick and cheap landing page.

Zoho.com- Zoho.com is my favorite CRM tool. We use it in our business to manage all of our customers. They recently changed their "Free" version so that it has less functionality, but it is still a great tool if you don't have a CRM and want one for free. The upgraded paid version is also fantastic and it's very reasonable.

Aweber.com- If you are looking to integrate a simple email marketing tool with AutoResponder emails then this is the one that I recommend. You can create online web forms/opt-in forms in minutes and then integrate those with your autoresponder emails. And the form itself can easily be installed on your WordPress website. Prices for Aweber start at $19.95 a month.

Elance.com- The world is full of eager and affordable people who will take on your project for cheap. Whether you have a creative project, technical project, or just need admin help, you can go to Elance.com to find someone around the world that can help you.

OnlineJobs.Ph - We are big fans of using the Philippines to outsource some of your work. The people in the Philippines are incredibly hard-working, honest, and they speak great AMERICAN-style English. This website is showcases resumes of people in the Philippines who are looking to work with American businesses.

Wordpress.org- We've talked about the importance of WordPress and this is the official WordPress site. Go there to learn more.

SEO Pressor- This is a plugin that we have used in our business to help us with Search Engine Optimization. If you're looking to do your own SEO then I highly recommend that you invest in this plugin for your WordPress website.

Business Network International (BNI) - If you are a small business owner and are looking to increase the quality and quantity of your referral business then I HIGHLY recommend incorporating Business Network International (BNI) into your marketing strategy. BNI is a referral organization that we joined many years ago and it is one of the reasons why we started our business, WebsiteIn5Days.com! The structure of the organization, the quality of people you meet, and the fact that you have access to people all around the world through BNI is amazing and you it can help you dramatically grow your business. Go to wwwBNI.com to find a chapter near you.

Notes

Contact Information

As I mentioned a couple times in the book, if you have a proposal that you want us to look at or you need some advice feel free to give me a call directly and I'll be more than happy to speak with you free of charge! Thanks so much for reading this book and starting your quest of online domination.

Chris Martinez
WebsiteIn5Days.com
2421 W 205th St
Suite D102
Torrance, CA 90501
Phone: 1-800-935-3168

Your $497 Free Offer Just For Investing In This Book!!

As promised, we are giving you a Free Offer worth $497. Take all or pick and choose which giveaways you want to take advantage of. It's totally up to you!

First, you can get a custom website evaluation from WebsiteIn5Days.com absolutely FREE. This is a $199 value!

Next, we will also give you (4) Blog Articles that you can use on your website ($199 Value)! And finally, we will also create an SEO Keyword Report for your business that will tell you what are good keywords to target so you can get more traffic and more customers a ($99 Value)!

Call us at 1-800-935-3168 or visit www.WebsiteIn5Days.com/wi5d497 to take advantage of these amazing free offers!